COUNTDOWN TO SPACE

INTERNATIONAL SPACE STATION
A Space Mission

Michael D. Cole

Series Advisor:
John E. McLeaish
Chief, Public Information Office, retired,
NASA Johnson Space Center

Enslow Publishers, Inc.

44 Fadem Road	PO Box 38
Box 699	Aldershot
Springfield, NJ 07081	Hants GU12 6BP
USA	UK

http://www.enslow.com

Copyright ©1999 by Enslow Publishers, Inc.

Library of Congress Cataloging-in-Publication Data

Cole, Michael D.
   International space station : a space mission / by Michael D. Cole.
      p. cm.— (Countdown to space)
   Includes bibliographical references and index.
   Summary: Discusses the need for the proposed international space station, its
design, and purpose, based partly on experiences with space station Mir.
   ISBN 0-7660-1117-8
   1. Space stations—Juvenile literature. 2. Astronautics—International
cooperation—Juvenile literature. [1. Space stations. 2. International
cooperation.] I. Title. II. Series.
TL797.C625   1999
629.44'2—dc21   98-18781
                                                 CIP
                                               AC

Printed in the United States of America

10 9 8 7 6 5 4 3 2 1

To Our Readers: All Internet addresses in this book were active and appropriate
when we went to press. Any comments or suggestions can be sent by e-mail to
Comments@enslow.com or to the address on the back cover.

Illustration Credits: National Aeronautics and Space Administration
(NASA)

Cover Illustration: NASA (foreground); Raghvendra Sahai and John
Trauger (JPL), the WFPC2 science team, NASA, and AURA/STSCI
(background).

CONTENTS

The International Space Station will be a permanent place in space where astronauts can live and work.

1

Danger at Space Station Mir

Orbiting 190 miles above Earth was the Russian space station *Mir*. Russian cosmonauts Vasili Tsibliyev and Alexander Lazutkin and American astronaut Michael Foale were aboard the station. A difficult task awaited them.

A Russian *Progress 1* resupply vehicle was to be docked to the space station the next day. Using remote controls aboard the station, without any help from the automatic guidance system the Russian space agency had once used, Commander Tsibliyev was expected to steer the vehicle into position and dock it. The job would not be easy, and *Mir*'s commander knew it.

"This is a bad business," Tsibliyev said to Lazutkin

as they ate their dinner. "It is bad. It's a dangerous thing to do."[1]

Lazutkin tried to assure Tsibliyev that it would be okay. Foale, who was new to the station, just listened.

On the following day, June 25, 1997, the three men took their positions for the docking. They were ready. Tsibliyev used remote joysticks to control *Progress 1* and bring the vehicle closer to *Mir*. When *Progress 1* was about thirty-five hundred feet away, Tsibliyev noticed something was wrong. The vehicle was coming in too fast.

He grabbed the joysticks to provide a braking thrust to *Progress 1*. But the vehicle kept moving toward *Mir*. It was not stopping! Tsibliyev hit the joysticks hard to give the vehicle's thrusters another braking burst, but it kept coming toward them.

"Michael," Tsibliyev said to Foale, "try getting a range mark."

Foale looked through the module's small window and could not yet see *Progress 1*. "Nothing," he said. He could not see the vehicle because it was approaching *Mir* from above.

Lazutkin also could not see it. Yet *Progress 1* rapidly closed to within 150 feet from the station.

"Try to get another range!" Tsibliyev shouted to his crewmates.

When Lazutkin looked out his window again, he was

One month prior to their trouble on Mir, the crew posed for a group photograph with a space shuttle crew. The three Mir crew members in blue suits are (from left to right) Russian cosmonauts Vasili Tsibliyev and Alexander Lazutkin and American astronaut Michael Foale.

stunned to see *Progress 1* passing over almost on top of them.

"There it is already!" he shouted. "It's coming in! Fast!"

Looking at the television monitor that enabled him to see the approaching *Progress 1*, Tsibliyev could tell that it was going to hit the station.

"Michael, get into the spacecraft!" he shouted to Foale.[2]

Tsibliyev was asking Foale to hurry from the module into the Soyuz spacecraft, which was docked to *Mir* in case an emergency caused them to abandon the station. This was definitely an emergency.

"Until the very end I was holding the handles to try and get the craft not to hit the station," Tsibliyev said. "If it had hit us directly, it would have punctured the core module directly and we would have all died."[3]

Before Foale reached the Soyuz, he heard the vehicle crash into *Mir*. A loud alarm went off inside the station. A popping sensation inside his ears told him that air pressure was slowly leaking from the space station.

Immediately he began to disconnect all the cables to the Soyuz craft so that they could eventually close its hatch. Suddenly Lazutkin was beside him. They were in a part of *Mir* that served as a connecting point for the Soyuz and four other parts of the station. Instead of helping Foale, Lazutkin hurriedly began disconnecting the cables to the Spektr science module. He had seen *Progress 1* hit Spektr and was trying to seal it off.

A half hour later, Lazutkin and Foale had disconnected all the cables. They were ready to slam the Spektr hatch shut. Shortly after the hatch closed, the pressure inside *Mir* stabilized. But their troubles were not over.

The impact of *Progress 1* had knocked the station into a slow roll and had caused a loss of power. The instrument panels were dark. Unable to read their instruments, the three men could not control the station.

The station's rolling had to be stopped. If they could determine the rate at which *Mir* was rolling, the ground controllers could fire *Mir*'s thrusters and stabilize the station. But they had no instruments working to help them determine the roll.

Astronaut Foale knew a little trick he thought might work. He went to one of the station's windows and held out his thumb at arm's length. Holding his thumb this way, Foale knew he could cover an area of sky that was about one and a half degrees of arc. Watching the stars come and go behind his thumb while checking his watch, Foale got a close estimate of how fast the station was rolling. After several minutes, he believed he had it.

"Tell them we're moving one degree per second," Foale said.[4]

His estimate was radioed to the Russian ground controllers. The controllers later fired *Mir*'s thrusters to stop a one-degree-per-second roll. Moments later the station stopped rolling. The sunlight hit the solar arrays, and *Mir*'s instruments came back to life. Tsibliyev, Lazutkin, and Foale were saved.

Not until later did Foale think about the danger he had been in.

"You don't get frightened," he said, "you just kind of go into this mechanical mode of thinking things through and trying to figure out what to do next quickly. Days

On June 25, 1997, a Progress resupply vehicle crashed into the Mir space station. The damaged solar panel (right) and radiator caused the module to depressurize.

later, when I thought about what had happened, that's when I got a little bit frightened by it all."[5]

With intelligence, teamwork, and the help of their ground controllers, the three men had saved themselves and space station *Mir* from what might have been a terrible disaster.

Intelligence and teamwork in every area of the project is essential to the success of a space station. The design, construction, and function of an orbiting space station is an amazingly complex operation. Through the end of the 1990s, the space agencies of Russia, the United States, and other countries continued to use *Mir* as an experimental laboratory in space. It helped prepare them for the task that lay ahead: the construction of a newer, larger, and better orbiting station in space, the International Space Station.

2

The Need for an International Space Station

Why do we need an international space station? If we believe that part of our future lies in space, then we must have a permanent place in space in which to work.

The International Space Station (ISS) will allow men and women to live in habitation modules and work in various laboratory modules for four to five months at a time, sometimes even longer. Having people stay in space for these periods will tell scientists more about the effects of weightlessness on the human body. For example, if we plan to send people on a ten-month journey to Mars, then we must have many people spend ten or more months in space to see how the experience affects them.

The new station will provide the most advanced

facilities in which to conduct many kinds of scientific experiments. The ISS will have more room in which to live and work than there has ever been in space before. Scientific research aboard the station will not only assist our efforts to continue exploring space, it may also lead to exciting discoveries that will improve our lives on Earth.

Studies of the combustion process (explosions and the burning of fuels) could lead to energy savings on Earth. Learning how to burn fuel more efficiently will save on energy bills for millions of people and decrease the amount of pollution in the air.

Laboratory and habitation modules of the International Space Station will allow scientists to live and work in space.

Work on the development of new silicon crystals may lead to the production of more powerful microchips for computers. The ability to design more powerful computers will greatly aid humans in their exploration of space and in their daily lives on Earth.

Studies of Earth's atmosphere and observations of its geologic formations and climate changes can also be conducted from the station. Because it orbits above Earth's atmosphere, the station will be an excellent place to conduct astronomical studies of the Moon, the planets, and the stars.

Studies of protein growth in the weightlessness of space, or microgravity, will increase our understanding of proteins, enzymes, and viruses. This increased knowledge may help scientists to develop new drugs and vaccines to fight a number of diseases.[1]

Without the effect of Earth's gravity, microgravity research can be done on the ISS. Studies in how the human body works give important knowledge on how our bones develop and grow. It gives new insight on how our sense of balance works and how it can be improved. The findings from these studies will prove very important when humans make long journeys through space to Mars and other planets.[2]

The weightless environment of space also affects the development of materials in many unusual and useful ways. Experiments in the station's laboratories may show scientists how to develop lighter, stronger metals. Such

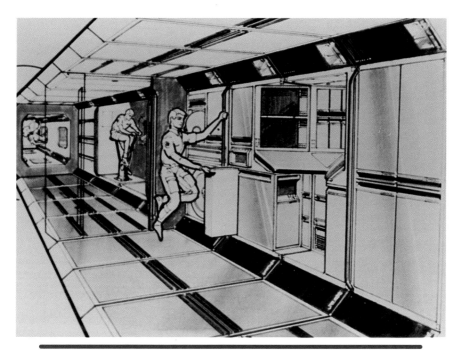

This illustration of a habitation module has an area with a rowing machine and other exercise equipment. In the background there is a zone that contains a shower, a lavatory, and station monitoring equipment.

discoveries would lead to stronger building materials for houses and skyscrapers. Lighter and more durable metals for aircraft, cars, and even wheelchairs may be created. Earth's gravity does not allow such materials to be developed on the ground.[3]

Knowledge gained from studies aboard the station may indeed come down to Earth to affect your life someday. You or someone you know may be saved from a dreadful illness by a special drug that was developed in the station's microgravity. With the addition of a microchip made from the silicon crystals developed aboard the station, your computer may run better and faster than you ever could have imagined.

It will take a large and busy space station to accomplish these goals. That is why plans for a space station have existed since before humans ever went into space.

The space programs for both the United States and Russia began in the late 1950s. By 1961, both countries had successfully launched men into space in small spacecraft attached to large rockets. The used rockets fell away to crash into the ocean or to burn up while falling back through Earth's atmosphere. Only the small spacecraft came back to Earth in one piece. These early spacecraft were not very roomy at all. They were only large enough to hold the astronaut safely inside.

Larger spacecraft that held two to three astronauts followed. On all these early flights, including the United States trips to the Moon, astronauts conducted various scientific experiments within the cramped spacecraft. There was much to learn about space and space travel.

To conduct more scientific experiments and further study how humans reacted to the effects of weightlessness, a larger spacecraft was needed. Astronauts needed a craft with more room in which to work and move around. They needed a craft that was full of equipment they could use to conduct experiments and to study their space environment.

They needed a space station.

Russia, formerly part of the Soviet Union, launched and manned the first space station, *Salyut 1*, in 1971.[4]

The *Soyuz 11* crew who docked with the *Salyut 1* worked aboard the station for twenty-three days. They conducted numerous experiments. This first mission aboard the station was a triumph for the Russian space program. Unfortunately this triumph ended in tragedy.

The *Soyuz 11* crew of V. I. Patsayev, G. T. Dobrovolsky, and V. N. Volkov was killed when a valve accidentally opened during the return to Earth. The opened valve allowed the life-sustaining atmosphere inside their Soyuz spacecraft to leak into space. The spacecraft's parachutes opened automatically and the spacecraft floated down to a landing in Russia. The recovery crew who opened the spacecraft was stunned to find all three cosmonauts dead in their seats. The accident caused Russian manned spaceflights to be grounded for more than two years.[5]

At the end of that period, nine different Salyut stations were launched through the 1970s. It was not until *Salyut 6* that the program began to be successful and productive.

On May 14, 1973, the United States launched its own space station laboratory, *Skylab*. It included a telescope, an exercise bike, and an airlock for going outside the station into space. Just over a minute after liftoff, *Skylab* was damaged when the aluminum sheeting around the station was torn away, damaging two of its folded solar arrays. Although *Skylab* achieved orbit successfully, it was crippled until its problems could be solved.

When the first three-astronaut crew arrived at *Skylab* aboard an Apollo spacecraft more than a week later, they immediately went to work on the problems. After a series of space walks, they had corrected the problems enough to safely enter *Skylab*.[6]

The repairs worked well enough for three separate three-man crews to live and work aboard the station. They completed stays of twenty-eight, fifty-nine, and eighty-four days aboard *Skylab*, before the last crew returned to Earth in February 1974.

Skylab and the Salyut stations were all single, large-laboratory units. They were fitted with docking ports for spacecraft designed to carry astronauts back and forth from Earth. The next Russian space station, *Mir*, was something new and different.

Mir became the first modular space station, meaning that it was composed of a number of different units. Docking and sealing these units, or modules, together constructed a larger station.

The *Mir* core module, launched on February 20, 1986, was essentially an enlarged, improved version of the Salyuts. The nose of *Mir* had a multiple docking adapter, with one forward docking port and four additional ports. If you were aboard *Mir* and facing forward in the docking adapter, there would be one docking port

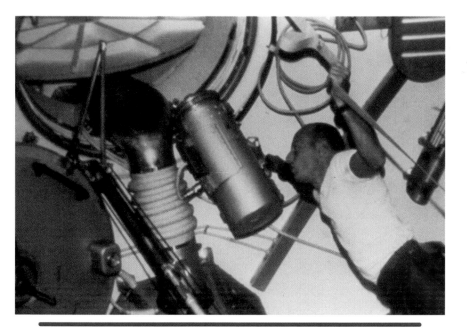

Specially designed telescopes were used on Skylab to study and photograph our sun from above the Earth's atmosphere.

directly ahead, one to the right, one to the left, one facing down, and one directly overhead. There was also a single docking port at the rear of the *Mir* core module.

The first Russian crews arriving at *Mir* in 1986 docked their spacecraft to the forward port. The spacecraft and the station faced each other nose to nose. In the years that followed, laboratory modules were launched into space and attached to the other five ports around the docking adapter. By 1996, all six docking ports were occupied.

The station has been permanently manned since September 1989. More than eighty different spacecraft have visited *Mir*, carrying astronauts, supplies, and equipment to and from the station. At first, only Russian cosmonauts visited the station, usually for a period of two to four months. In the early 1990s, Russia began to carry astronauts from the European Space Agency (ESA) to the station. The ESA handles the space efforts of member European countries, such as France, Germany, Italy, and Great Britain.

When the Soviet Union fell in 1991, both Russia and the United States decided it was time to try to work together in space once again. (Years earlier, in 1975, a United States Apollo spacecraft and a Russian Soyuz spacecraft had docked in space.)

In March of 1995, United States astronaut Norm Thagard, together with two Russian cosmonauts, lifted off in a Soyuz rocket. The rocket blasted off from the Baikonur Cosmodrome in the

The space shuttle Discovery *took this photograph of* Mir *while it orbited the earth above the Pacific Ocean.*

former Soviet country of Kazakhstan. They were bound for *Mir*. The next day Thagard became the first American astronaut to go aboard a Russian space station.[7] On June 29, 1995, the American space shuttle *Atlantis*, carrying five United States astronauts and two Russian cosmonauts, became the first United States spacecraft to dock with the Russian space station.

Some time after the docking, commander Robert "Hoot" Gibson opened the hatch and climbed through the airlock to shake hands with *Mir* commander

Shuttle astronaut Robert Gibson (in red) offers a smile and a handshake to Mir *cosmonaut Vladimir Dezhurov after* Atlantis *and* Mir *linked in space.*

Vladimir Dezhurov in the docking tunnel. Russian and American astronaut crews were joined in orbit again. A short time later the seven shuttle crew members joined *Mir's* three crew members, including American Norm Thagard. They had a welcoming ceremony and an exchange of gifts in the station's core module.

For five days, the crews worked on different experiments. Most of these experiments were focused on how Thagard, Dezhurov, and cosmonaut Gennady Strekalov were reacting to their nearly four-month stay aboard the orbiting station.

When *Atlantis* was ready to depart, the three original *Mir* crew members and the five astronauts from *Atlantis* climbed through the docking tunnel and boarded the shuttle to return to Earth.[8] Never before had there been so much international cooperation involved in a single space mission. Shortly after *Atlantis* touched down on the runway at the Kennedy Space Center in Florida, President Bill Clinton radioed the *Atlantis* crew.

"This is truly the beginning of a new era of cooperation in space between the United States and Russia," President Clinton told them. "Because of your mission, the United States and Russia, with our partners in Canada and Japan and Europe, are going to be able to meet the challenge of building the International Space Station."[9]

3

Building the Space Station

Even before the United States-Russian missions to *Mir*, the space agencies of the United States and Russia, along with those of other countries, were putting together plans for the International Space Station. Other space stations had been planned in the past, but the high cost of building them kept them from getting off the drawing board. Space stations *Freedom* and *Alpha* were two that were never built. Their designs, however, contributed much to the present plans for the ISS.[1]

Cooperation between a number of countries has made the space station possible. Dividing the cost of building the station among these countries made the monetary burden lighter for all. The United States, Russia, Canada, Japan, Belgium, Denmark, France,

Germany, Italy, the Netherlands, Norway, Spain, and the United Kingdom have all contributed to the ISS project.

The plans for operating and constructing the new station will have been completed in three main stages. The first stage ran between 1994 and 1998. This stage consisted of the cooperative missions between the United States and Russia using the space shuttle and the *Mir* space station. The goal of these missions was to build technical experience in areas such as space station construction. Further experience was also gained in how to successfully conduct cooperative international space missions. Some space walks conducted from the space shuttle tested tools and materials astronauts will use in the construction of the ISS.

In 1996 and 1997, *Mir* began to experience an increasingly frequent number of problems and technical failures. Weeks after the collision between the *Progress 1* resupply vehicle and the station's Spektr module, a key computer cable was accidentally unplugged, temporarily cutting *Mir*'s power supply. In August and September of 1997, the station's oxygen generators broke and had to be repaired. During the same period, *Mir*'s main onboard computer failed three times. It had to be replaced with a new main computer. Solar arrays were repaired, and one was totally replaced. There were further problems with docking.[2]

Part of these problems were due to the fact that *Mir* was originally designed for a seven- to eight-year mission

The completed International Space Station will be 262 feet long, 356 feet wide, and weigh more than one million pounds.

in orbit. By 1998, the aging space station had been in orbit for twelve years, far past its intended length of service. Yet the astronauts and cosmonauts managed to solve the frequent problems with *Mir,* providing crew members and space agencies with an unexpected benefit—additional experience in spacewalking, international space communications, and general problem-solving in space. Frank Culbertson, NASA's manager of the Shuttle-*Mir* program, believed the experiences were valuable.

"I believe what's going to come out of this is a better

understanding, not only of the real risks of operating in space," he said, "but a better understanding of what it takes to manage those risks and to deal with them when something unexpected happens."[3]

Stage two of the plan marked the beginning of actual construction in space of the International Space Station. This stage is scheduled to run between mid-1998 and late 1999. It will begin with the launch aboard a Russian Proton rocket of the module called the Functional Cargo Block. It is referred to by its Russian acronym, FGB, and is named Zarya, the Russian word for "sunrise."[4]

The FGB will provide power and orbital control to the station in its early stages of construction. It is over forty feet long and more than thirteen feet in diameter. Weighing twenty-one tons, it is one of the largest modules of the station. The FGB will control the motion of the station and maintain the altitude of the station's orbit. In the station's later operation, the FGB will primarily be used for fuel storage. It will have a docking port at one end for connection to the Russian Service Module and another port at the opposite end for connection to the United States Node 1 module, named Unity.

The Russian Service Module will provide life-support systems to the early station and additional orbit-boosting capabilities.[5] It will also provide the astronaut living quarters. Similar to the *Mir* core module, it will include a work compartment, sleeping quarters, a

Russian technicians finish work on the Functional Cargo Block (FGB). This section will be the first launched part of the International Space Station.

treadmill and exercise bike, and a transfer compartment that can be used as an airlock. The rear of the module will have a docking port that will be used for connection to Russian Progress resupply vehicles. These unmanned Progress vehicles will be the primary resupply ships for the station.[6]

The Unity module will serve as a major connecting passageway for the station. One of the module's six hatches will connect to the FGB, with the opposite hatch leading to the United States Laboratory Module. The remaining four hatches of Unity will provide connecting points to other future components of the station. These will include the United States Habitation Module, an airlock, and the truss structure.

The truss structure is the long metal beams that will span the station from one end to the other. These structures will be like a backbone for the station and will connect the station's giant solar arrays.

Months after the FGB, the Russian Service Module, and Unity are connected in space, a Russian Soyuz spacecraft will arrive with the station's first three-person crew. They will spend five months at the space station. Although the completed station will be able to accommodate a crew of up to seven, the early stages of the station will be limited to a crew of three. The space shuttle will carry the station's later crew members to and from the station and will continue to deliver further building components. Permanent occupation of the station, however, will begin with the docking of Soyuz. When no space shuttle is present at the station, the docked Soyuz spacecraft will serve the station's crew as its Assured Crew Return Vehicle.[7]

Construction will continue as the space shuttle delivers the space station's Remote Manipulator System (RMS). This large robotic arm is the Canadian Space Agency's contribution to the ISS. It will include the Canada Hand, which can be attached to the end of the arm. The Canada Hand is like a set of large robotic fingers that can handle delicate assembly tasks. These tasks used to only be done by attaching spacewalking astronauts to the end of the arm.

The RMS will be an important tool of the space station. It is capable of moving large payloads and assisting in docking the shuttle. It will become a useful tool in the space station's construction. The RMS will also aid astronauts in repairing and servicing satellites and other spacecraft brought to the station. The RMS arm is secured to its movable base by a special latch. The latch will allow the arm to be removed and attached to

This forty-five foot section of the main beam of the International Space Station was assembled in one hour. The astronaut is standing on the end of the Canada Arm, showing how easy it is to move the materials in microgravity even while wearing a bulky space suit.

several RMS port locations around the outside of the station. There it can assist in other docking or construction tasks.[8]

Delivery to the station of the United States Laboratory Module will bring the second stage of the space station's construction to a close.

The third stage will mark the beginning of the International Space Station's continuous occupation by a full crew of seven. Construction of the station will continue, but not all of the United States and Russian flights to the station during this period will deal with the station's construction. Some will be devoted solely to bringing supplies and scientific equipment to the station for use in the crew members' onboard experiments. Another Soyuz spacecraft will also be docked to the station to serve as a second "lifeboat" to the seven-member crew.

Construction during this period will include installation of the Japanese Experiment Module and the European Space Agency's Columbus Orbital Facility (COF). The COF is a laboratory for microgravity experiments in material sciences, fluid physics, and biology. It will be used by all astronauts from Europe, the United States, Russia, Japan, and Canada.[9]

The final phase of the International Space Station's construction will come to a close with the addition of the station's largest module, the United States Habitation Module. This module will serve as one of the main living,

The X-38 crew return vehicle can be seen under the wing of the B-52 aircraft during drop tests. A return vehicle will, of course, be a very important part of returning astronauts safely from the ISS.

sleeping, and recreational quarters for the station's crew members. It includes an additional kitchen, a toilet, a shower, sleep stations, and medical facilities.

Like most other living and working environments on the station, the United States Habitation Module is designed in a normal floor-to-ceiling arrangement. This means that the module's tables and chairs are attached to only one surface of the room. Equipment and instrument panels are arranged so that crew members can work at them as if they were standing normally on the floor—the same floor to which the table and chairs are attached.

Because there is no real up or down in space, there is no need for one surface to be designated as the floor and another as the ceiling. *Skylab*, for example, had instruments, chairs, and exercise equipment attached to every available surface of the station's tube-shaped interior. This proved to be a productive and successful arrangement.

Yet astronauts found the absence of an identifiable floor and ceiling inside *Skylab* confusing. They could work in such an environment, but they found it far less comfortable than the floor-to-ceiling arrangement they

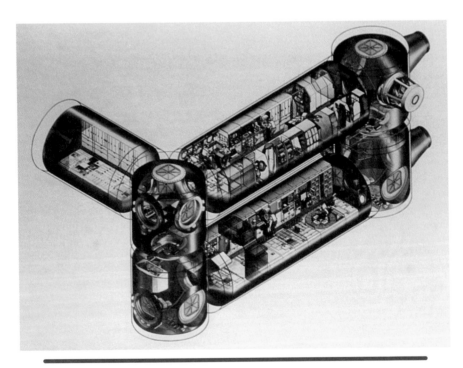

Each of the large modules on the International Space Station will be about the size of a house trailer.

were accustomed to on Earth. That is why most of the modules on the ISS have been designed in a floor-to-ceiling arrangement. It allows the station's crew members to live and work in a more comfortable or "normal" environment while spending long weeks or months in orbit.

Completion of the ISS will take nearly thirty space shuttle flights and more than forty Russian flights.[10] The station's construction will cost the countries involved in the project more than sixty billion dollars.

In its completed form, the station will consist of six main laboratories, two main living quarters, two Soyuz crew transfer vehicles, and a number of other units that will provide engine thrust and electrical power to the station.[11]

Design and planning for the station began as far back as the early 1980s. It will have taken more than twenty years to move the project from plan to reality. Building the International Space Station will require a tremendous effort not only by astronauts, but by tens of thousands of scientists, engineers, and other workers in countries all over the world.

4

Our Future with the ISS

Sometime after the year 2000, a mission like the following one may take place.

An American space shuttle rumbles off the launchpad in Cape Canaveral, Florida. Its engines and solid rocket boosters trail a column of bright fire and billowing smoke as it thunders into the sky toward space. A few minutes into the bumpy flight, the ride smoothes out as the solid rocket boosters detach from the shuttle and fall away. The astronauts look out the window at the Earth below and the blackness of space above. They are in orbit.

Most of this astronaut crew has flown in the shuttle before. Three of the astronauts are American, two are Japanese, one is Russian, and another French. The

Russian and one of the Americans have already been to where they are going—the International Space Station.

After several orbits in which the astronauts adjust to weightlessness, the pilot guides the shuttle toward a docking with the space station. As the shuttle draws closer to the station, the astronauts watch out the windows as their view of the station grows larger.

The shuttle's payload bay doors are already open. The pilot moves the shuttle into position so that the airlock in the open payload bay is facing the station's docking port. Using the shuttle's thruster controls, the pilot carefully guides the shuttle's airlock toward the docking port.

The first crew to visit the completed space station will blast into space aboard a space shuttle.

The pilot gives the shuttle one burst from the thrusters, then another, and another. Finally, the airlock clamps against the docking port. The shuttle is docked to the ISS.

Since its first steps of construction in the late 1990s, the space station has become a huge and complex structure. One of the astronauts on the shuttle had helped construct part of the station just a few years earlier. Now the ISS is fully operational. .

A crew of six astronauts has already been aboard the space station for several weeks. Two of those astronauts will be relieved by the French and one of the Japanese astronauts aboard the shuttle.

The French astronaut is a microbiologist. He has come to the station to conduct medical research experiments in one of the station's laboratories. The Japanese astronaut is an engineer. She will conduct experiments on different metals by melting them down and seeing how they solidify in microgravity. This may lead to the development of stronger metals for use in construction on Earth.

The rest of the crew will be delivering a cargo of new supplies and special equipment to the station. The six astronauts who remain behind on the station will need these supplies to continue living and working in space for the next few weeks or months.

A big job awaits the rest of the shuttle crew while they are docked.

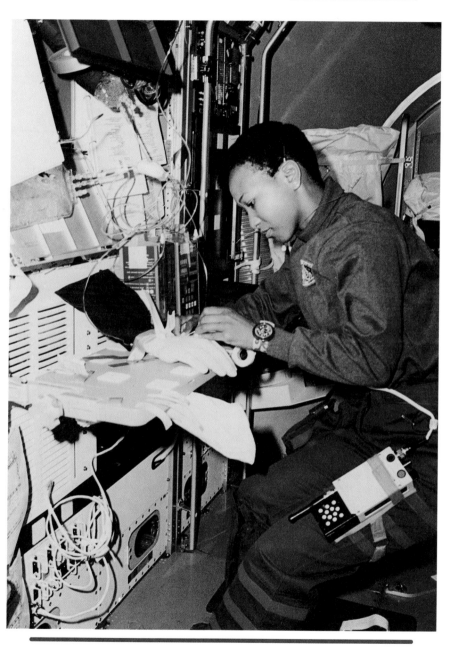

The fully operational space station will allow astronauts to conduct research in space.

A previous shuttle mission had snatched a malfunctioning communications satellite from its orbit. After the shuttle had docked with the space station, the satellite was removed from the shuttle's payload bay with the station's RMS. The RMS had then moved along the station's backbone of metal trusses, transporting the satellite to a storage area, where it had been secured.

Some time after that shuttle departed, astronauts from the station had ventured outside in their space suits

The RMS on the space station will be used to grab satellites that may need repair. The arm of a space shuttle's RMS can be seen at the top of the photograph, holding this large satellite.

to remove several components from the satellite. These large components had been placed on other storage points along the metal trusses with the help of the RMS.

The job of the current shuttle astronauts is to replace those large components with new ones they have brought aboard the shuttle's payload bay. With the help of the station's RMS, these astronauts are to conduct a series of space walks to place the large new components into the satellite.

Eventually the difficult space walks are completed and the new components are safely inside the satellite. But the work of the spacewalkers and the station's RMS operator is only half over. The satellite's older components are still attached to the station. The spacewalkers and the RMS operator must once again work together to move the older components into the shuttle's payload bay, where they will be stowed and carried back to Earth. The space walks take hours and do not always go exactly as planned.

During one of them, a small screw slips from one of the astronauts' hands. This escaped screw, despite being very small, poses no minor threat. The floating screw presents a danger.

The astronaut at the end of the RMS arm tells the RMS operator in the station to move him in the direction of the slowly spinning screw. If the screw becomes trapped somewhere amid the complex hardware of the station or the shuttle, it could cause serious damage to components and may threaten the lives of the crew. Secured to the end of the RMS arm, the astronaut chases the slowly tumbling screw.

"Steady . . . steady," he tells the RMS operator over his helmet headset. "A little more to the left." He reaches out as the RMS operator warns him that the arm is nearing its full extension. He must grab the screw before it is lost, perhaps becoming dangerously lodged around the docking bays or airlocks. He himself is about to collide with the station's metal truss, which could tear a hole in his suit.

Reaching out as the RMS arm guides him closer to the truss, the astronaut grasps the screw in his glove.

"Arm stop!" he quickly says, and the RMS operator immediately halts the arm. The danger of losing the screw is avoided, and there is a quiet sigh of relief as the astronauts continue their work.[1]

Brief emergencies such as this will be, to some degree, unavoidable. While the dangers of space can never be totally eliminated, the ISS will make living and working in space much easier. Scientists and engineers working in the laboratories aboard the station will have more room and better equipment than there has ever been in space before. Their work should produce important knowledge that can be used for space exploration and in our daily lives.

On clear nights, young people watching the night sky will be able to see the bright glint of light from the space

Although the hazards of working in space can never be totally eliminated, the International Space Station will make life in space much easier.

station as, high above them, it orbits Earth. That bright glint of light is likely to inspire many young people to make the excitement of space a part of their own future.

Imagine that sometime between the year 2020 and 2030 you are selected as a crew member for a voyage to Mars. Part of your training and preparation will take place in space. The many modules of your ship may need to be constructed in space where spacewalking astronauts can get to them. For your mission to be a success, you will need the many facilities of a space station to prepare.

But it will not end there. It is very likely that many of the components aboard your spacecraft owe part of their development to research conducted on the space station. Experiences gained aboard the station have already contributed a great deal to the design and construction of your spacecraft and to the planning and possible success of your mission. In fact, it is difficult to imagine how your mission, and many others like it, would ever have been possible without the space station.

Maybe living and working in space really *is* in your future. If so, you might look forward to spending some very interesting times in orbit aboard the International Space Station.

CHAPTER NOTES

Chapter 1. Danger at Space Station Mir

1. Jeffrey Kluger, "A Bad Day in Space," *Time*, vol. 150, no. 18, November 3, 1997, p. 86.

2. Ibid., p. 89.

3. Jim Banke, "Moscow, We Have a Problem," *Ad Astra*, September/October 1997, p. 33.

4. Kluger, p. 91.

5. Banke, p. 34.

Chapter 2. The Need for an International Space Station

1. Hugh Ronalds and Terry McDonald, "Scientific Disciplines," *NASA International Space Station Home Page*, December 4, 1997, <http://station.nasa.gov/sreference/factbook/med.html> (March 22, 1998).

2. Ibid.

3. Ibid.

4. Phillip Clark, *The Soviet Manned Space Program* (New York: Salamander Books, 1988), pp. 58–63.

5. Ibid., pp. 64–65.

6. John and Nancy Dewaard, *History of NASA: America's Voyage to the Stars* (Greenwich, Conn.: Brompton Books, 1984), pp. 144–159.

7. Jim Banke, "The United States and Russia Begin a New Partnership in Space, But Can We Trust Each Other?" *Ad Astra*, vol. 7, no. 5, May/June 1995, p. 23.

8. Jim Banke, "Shuttle's Summer Successes," *Ad Astra*, September/October 1995, p. 18.

9. Ibid.

Chapter 3. Building the Space Station

1. Committee on Space Stations Report: *The Capabilities of Space Stations*, National Research Council—Aeronautics and Space Engineering Board, Commission on Engineering and Technical Systems (1995), Chapter 1, p. 1; Chapter 5, p. 1.

2. "A Troubled Year for *Mir*," uncredited article in *Launchspace*, February/March 1998, p. 19.

3. Jim Banke, "Moscow, We Have a Problem," *Ad Astra*, September/October 1997, p. 35.

4. "New Space Station Assembly Sequence Announced," uncredited article in *Ad Astra*, July/August 1997, pp. 14–15.

5. Ibid.

6. Committee on Space Stations Report, Chapter 5, p. 1.

7. "New Space Station Assembly Sequence Announced," pp. 14–15.

8. NASA Johnson Space Center, "ISS Technical Data Book," *Extravehicular Robotics Homepage*, 1995, <http://iss-www.jsc.nasa.gov/ss/techdata/EVR/EVR.html> (March 22, 1998).

9. ESA, *Manned Spaceflight and Microgravity Home Page*, February 1996, <http://www.esrin.esa.it/esa/progs/msm.html> (August 7, 1998).

10. NASA Marshall Spaceflight Center, *International Space Station Fact Book*, June 1, 1995, <http://spacelink.msfc.nasa.gov/t?NASA.Projects/Human.Space.Flight/Space.Station/International.Space.Station> (August 7, 1998).

11. Ibid.

Chapter 4. Our Future with the ISS

1. The fictional scenario described in the text was based on a real event documented in *Rescue Mission in Space: The Hubble Space Telescope*, NOVA Adventures in Science video (1994).

GLOSSARY

airlock—An airtight chamber that separates the pressurized interior of a spacecraft from the pressureless vacuum of space. Space-suited astronauts use airlocks to seal their spacecraft's pressure inside before they open an outer door to step into space. Without the airlock, the spacecraft's pressurized air would leak out the outer door into space.

enzymes—Proteins that cause reactions in the body.

microgravity—The sensation of weightlessness experienced by astronauts in orbit around Earth.

payload—The cargo and equipment that a vehicle, such as a space shuttle, carries inside it.

payload bay—The large area of the space shuttle that is used to carry cargo such as satellites, space station modules, and other equipment into orbit.

Remote Manipulator System (RMS)—The robotic arm that the space crews use to move large equipment around their spacecraft.

solar array—Silicon panels that use light from the sun to generate electrical power for spacecraft.

solid rocket boosters—The two rocket engines attached to the space shuttle's large external tank during liftoffs. The booster engines use explosive powders for fuel.

truss—The metal girders that serve as a connecting spine for several of the space station's components.

FURTHER READING

Books

Berliner, Don. *Living in Space*. Minneapolis: Lerner Publications Company, 1993.

————. *Our Future in Space*. Minneapolis: Lerner Publications Company, 1991.

Kettelkamp, Larry. *Living in Space*. New York: Morrow Junior Books, 1993.

Kramer, Barbara. *Sally Ride: A Space Biography*. Springfield, N.J.: Enslow Publishers, Inc., 1998.

Savage, Marshall T. *The Millennial Project: Colonizing the Galaxy in Eight Easy Steps*. Boston: Little, Brown & Company, 1994.

Internet Addresses

European Space Agency. *European Space Agency Microgravity Database*. n.d. <http://www.esrin.esa.it/esa/progs/msm. html> (August 7, 1998).

NASA. *International Space Station Home Page*. September 2, 1998. <http://station.nasa.gov> (September 3, 1998).

NASA. "Fact Book." *International Space Station*. May 22, 1998. <http://station.nasa.gov/reference/factbook> (September 3, 1998).

INDEX